Solids, Liquids, and Gases at the
Beach

by Fawn Bailey

Contents

Science Vocabulary

matter
Matter is anything that takes up space.

These rocks and boats are **matter.**

property
A **property** is something about an object that you can observe with your senses.

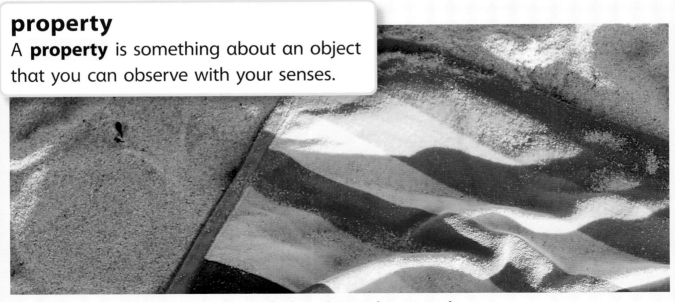

Softness is a **property** of this beach towel.

solid

A **solid** is matter that has its own shape.

These beach chairs are **solids.**

liquid

A **liquid** is matter that takes the shape of its container.

Water is a **liquid.**

gas

A **gas** is matter that spreads to fill a space.

Air is a **gas.** It fills this ball.

mixture

A **mixture** is two or more things put together.

This pitcher holds a **mixture** of fruit slices.

dissolve

When a solid **dissolves,** it mixes completely with a liquid.

solution

A **solution** is a mixture of two or more kinds of matter evenly spread out.

The salt **dissolves** in the water to form a **solution.**

volume

Volume is the amount of space matter takes up.

This person measures the water's **volume.**

My Science Vocabulary

dissolve

gas

liquid

matter

mixture

property

solid

solution

volume

water vapor

water vapor

Water vapor is a gas in the air.

water vapor

water

As the wet swimsuit dries, the water becomes **water vapor.**

Matter at the Beach

Matter is everywhere at the beach. Waves splash cold water on the sand and rocks. Boats sail on the ocean.

The boats, rocks, sand, and water take up space. You can't see the air, but the air takes up space, too.

matter
Matter is anything that takes up space.

Sand, rocks, boats, water, and air are matter. Anything that takes up space is matter.

Beach umbrellas shade people from the sun. A beach umbrella is a **solid.** The beach chairs are solids, too. The chairs and the umbrella each have their own shapes.

solid

A **solid** is matter that has its own shape.

People can fold their towels to fit them in their bags. But a seashell does not bend or fold. All matter does not change in the same way.

If you fold a beach towel, it will not break.

If you try to bend a seashell, it can break.

A beach ball sits on the sand. The beach ball is a solid. The inside is filled with air. Air is a **gas** that spreads out inside the ball and makes the ball round.

gas

A **gas** is matter that spreads to fill a space.

This person is drinking water. Water is a **liquid**. It doesn't have a shape of its own. The water takes the shape of the bottle.

liquid

A **liquid** is matter that takes the shape of its container.

Solids and Liquids Can Change

Don't drink the ocean water! Ocean water is salty. It is a **solution** of salt and water. The salt is spread out evenly in the ocean.

solution

A **solution** is a mixture of two or more kinds of matter evenly spread out.

You can make a smaller **volume** of saltwater at home. Volume is the amount of space matter takes up. Mix salt and water in a measuring cup. The salt **dissolves,** or mixes completely, with the water.

You can't see the salt after it dissolves.

volume

Volume is the amount of space matter takes up.

dissolve

When a solid **dissolves,** it mixes completely with a liquid.

People run, play, and splash in the ocean to cool off on a hot day. Swimsuits get wet from the ocean water.

Heat from the sun dries this swimsuit. The water becomes **water vapor.** It changes from a liquid to a gas and becomes part of the air. Gas fills the air outdoors, but you can't see it.

water vapor

water

water vapor

Water vapor is a gas in the air.

This beach has an ice cream stand. People can buy treats like frozen candy bars, ice cream cones, ice cream sandwiches, and ice pops. The freezer keeps the treats frozen and solid.

Ice will melt in the hot sun, though. This ice pop is melting fast. The sun's heat turns it into a liquid. Be careful! The liquid drips.

Observing Properties

People use their senses to observe **properties.**
A towel feels soft, but a pail feels hard. Texture
is a property. It is the way something feels.

property

A **property** is something about an object
that you can observe with your senses.

There are different temperatures at the beach. The sand under people's feet can be hot, and the water in the ocean can be cold.

These children run to the water to cool their feet and fill their pails.

There are lots of fun things to do at the beach. Some people make sandcastles at the beach. They build big sandcastles and small ones. Size is a property of matter.

Big

Small

People decorate their sandcastles with things they find along the shore. Feathers are light. Rocks are heavier. Weight is a property of matter.

This feather weighs less than these rocks.

Being in the sun can make people thirsty! Some make drinks with fruit slices. The pitcher is filled with a **mixture** of lemons and limes. A mixture is two or more things put together.

The fruit slices in this pitcher have different properties.

mixture

A **mixture** is two or more things put together.

The fruit slices have different colors and shapes. Color and shape are properties of matter. You can sort the fruit slices into groups by observing their different colors and shapes.

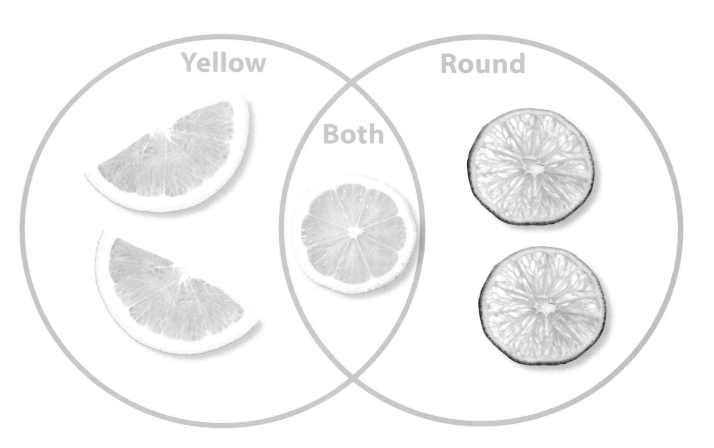

Some fruit slices are yellow, and some are round.
One slice is yellow *and* round.

In the ocean, some objects float and others sink. A boat floats in the water, but an anchor sinks to the bottom. Whether something floats or sinks is a property.

Matter at the beach has different properties that you can observe. How can you describe the matter at this beach? What solids, liquids, and gases can you find?

Conclusion

Solids, liquids, and gases are matter that you can find at the beach. Matter can change states. You can observe and measure the properties of matter at the beach.

Think About the Big Ideas

1. What are some solids at the beach?
2. What are some properties of beach towels that you can observe?
3. How can a seashell change if you try to bend it?

Share and Compare

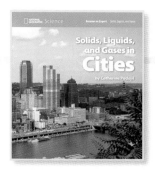

Turn and Talk

Compare the solids, liquids, and gases in your books. How are they alike? How are they different?

Read

Find a photo with a caption. Read the caption to a classmate.

Write

Describe the properties of a solid from your book. Share what you wrote with a classmate.

Draw

Draw a picture of matter from your book. Share your drawing with a classmate.

Meet Stephon Alexander

Stephon has always enjoyed exploring the world around him. When he was a child, he taught himself how to program old computers. He enjoyed making new video games on the computers.

Now, Stephon is a scientist who asks questions about space. He is curious about matter and where it comes from. He observes properties of matter to help him learn more about science.

Index

Acknowledgments
Grateful acknowledgment is given to the authors, artists, photographers, museums, publishers, and agents for permission to reprint copyrighted material. Every effort has been made to secure the appropriate permission. If any omissions have been made or if corrections are required, please contact the Publisher.

Photographic Credits
Cover (bg) Kraig Lieb/Lonely Planet Images; Cvr Flap (t), 5 (tl), 10 Ersler Dmitry/Shutterstock; Cvr Flap (c), 4 (b), 20 Teresa Hurst/iStockphoto; Cvr Flap (b), 19 Mira/Alamy Images; Title (bg) Polka Dot Images/Jupiterimages; 2-3 Christer Fredriksson/Lonely Planet Images; 4 (t), 8-9 James Schwabel/Panoramic Images; 5 (tr), 13 Image Source/Jupiterimages; 5 (b), 12 Gregory Olsen/iStockphoto; 6 (t), 11 (r), 24 Mark Thiessen and Becky Hale, National Geographic Photographers; 11 (l) Donall O Cleirigh/iStockphoto; 14-15 (bg), 28 Ron Chapple/Corbis; 16 Will Elwell/SuperStock; 18 Matt Baker/Alamy Images; 21 Corbis/SuperStock; 22 (l) En Tien Ou/iStockphoto, (r) druvo/iStockphoto; 23 (l) Sam Arnold/iStockphoto, (r) Photodisc/SuperStock; 25 (l) .AGA./Shutterstock, (c) Sasha Davas/Shutterstock, (r) Iconotec; 26 (l) Netfalls/Shutterstock, (r) Radius Images/Alamy Images; 27 Billy Stock/The Photolibrary Wales/Alamy Images; 31 Mark Thiessen/National Geographic Image Collection; Inside Back Cover (bg) Tischenko Irina/Shutterstock.

Illustrator Credits
6(b), 7(t), 15 (inset) Barbara Harmon; 7(b), 17 Sharon and Joel Harris

Neither the Publisher nor the authors shall be liable for any damage that may be caused or sustained or result from conducting any of the activities in this publication without specifically following instructions, undertaking the activities without proper supervision, or failing to comply with the cautions contained herein.

Program Authors
Malcolm B. Butler, Ph.D., Associate Professor of Science Education, University of South Florida, St. Petersburg, Florida; Judith Sweeney Lederman, Ph.D., Director of Teacher Education and Associate Professor of Science Education, Department of Mathematics and Science Education, Illinois Institute of Technology, Chicago, Illinois; Randy Bell, Ph.D., Associate Professor of Science Education, University of Virginia, Charlottesville, Virginia; Kathy Cabe Trundle, Ph.D., Associate Professor of Early Childhood Science Education, The Ohio State University, Columbus, Ohio; Nell K. Duke, Ed.D., Co-Director of the Literacy Achievement Research Center and Professor of Teacher Education and Educational Psychology, Michigan State University, East Lansing, Michigan; David W. Moore, Ph.D., Professor of Education, College of Teacher Education and Leadership, Arizona State University, Tempe, Arizona

The National Geographic Society
John M. Fahey, Jr., President & Chief Executive Officer
Gilbert M. Grosvenor, Chairman of the Board

National Geographic School Publishing
Hampton-Brown
www.NGSP.com

Printed in the USA.
Quad Graphics, Leominster, MA

ISBN: 978-0-7362-5600-1

18 19 20 21 22 23 24 25

10 9 8 7 6 5